A LITTLE BIT OF THIS DINOSAUR!

by Eileen Hutcheson
and Darcy Pattison
Illustrated by
John Joven

A Little Bit of THIS Dinosaur
by Elleen Hutcheson and Darcy Pattison
illustrated by John Joven

Text © 2023 Elleen Hutcheson and Darcy Pattison
Illustrations © 2023 Mims House, LLC

Mims House
1309 Broadway
Little Rock, AR 72202
USA

MimsHouseBooks.com

Publisher's Cataloging-in-Publication Data

Names: Hutcheson, Elleen, author. | Pattison, Darcy, author. | Joven, John, illustrator.
Title: A little bit of THIS dinosaur! / By Elleen Hutcheson and Darcy Pattison; illustrated by John Joven.
Description: Little Rock, AR: Mims House, 2022. | Summary: This humorous story follows a carbon atom as it journeys from dry bones to your spine.
Identifiers: LCCN 2022919194 | ISBN 9781629442242 (hardcover) | 9781629442259 (paperback) | 9781629442266 (ebook) | 9781629442273 (audio)
Subjects: LCSH Life cycles (Biology)--Juvenile literature. | Carbon in the body--Juvenile literature. | Bones--Juvenile literature. | Dinosaurs--Juvenile literature. | Humorous stories. | CYAC Life cycles (Biology). | Carbon in the body. | Bones. | Dinosaurs. | BISAC JUVENILE NONFICTION / Science & Nature / Biology | JUVENILE NONFICTION / Science & Nature / Anatomy & Physiology | JUVENILE NONFICTION / Science & Nature / Fossils | JUVENILE NONFICTION / Animals / Dinosaurs & Prehistoric Creatures
Classification: LCC QH501.H88 2022 | DDC 372.3/57--dc23

You have a little bit of Spinosaurus in your spine!

Don't believe me?
It's all your brother's fault.

Listen up.
Here's how
it happened.

Piu!
Piu!

Carbon Atom!

Once, in days of old, a volcano erupted, sending ash, smoke, and a carbon atom into the air.

The carbon floated on wind currents till it finally fell into a shallow river where it was absorbed by algae.

A hungry fish ate the algae.

A Spinosaurus chased the fish and caught it. Now the carbon atom that had spewed out of the volcano became a little bit of dinosaur.

When its days on earth were done, the spinosaurus died. Its bones hardened into fossils.

Time ticked by. The land dried and became a desert.

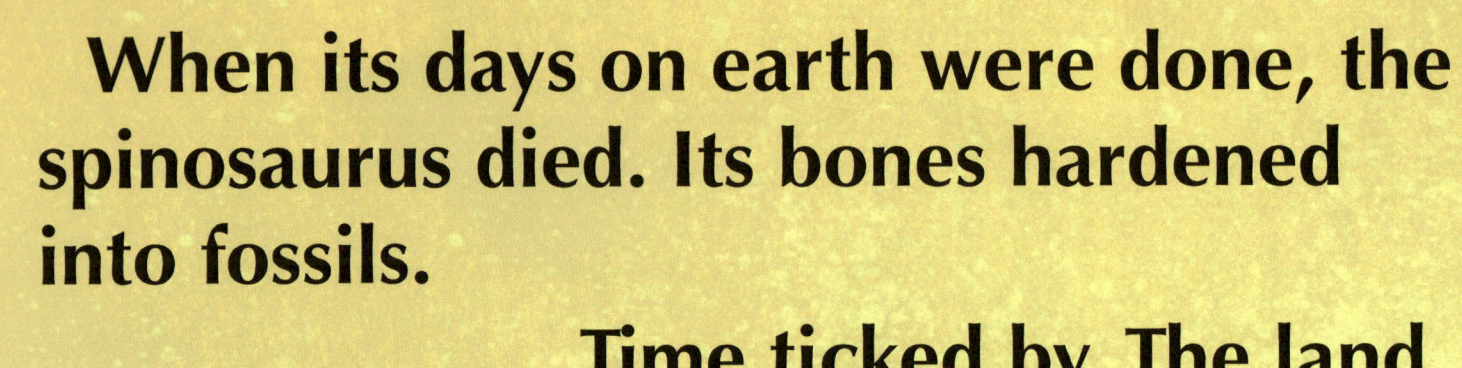

Sahara

Finally, in an oasis, a date palm grew near the old fossils. The roots wiggled close, touching the fossils. The palm tree took up the carbon. The little bit of carbon that used to be part of the Spinosaurus became part of a date.

A camel caravan stopped by the oasis, and a camel driver cut off all the stalks of dates.

At market, the camel driver
sold the dates.

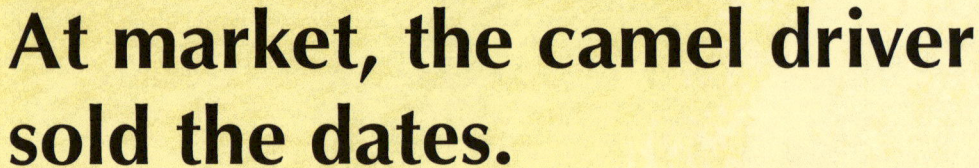

The dates were pitted and packaged and then flown to a warehouse.

A trucker delivered the dates to the store, and your father bought the dates for a family picnic.

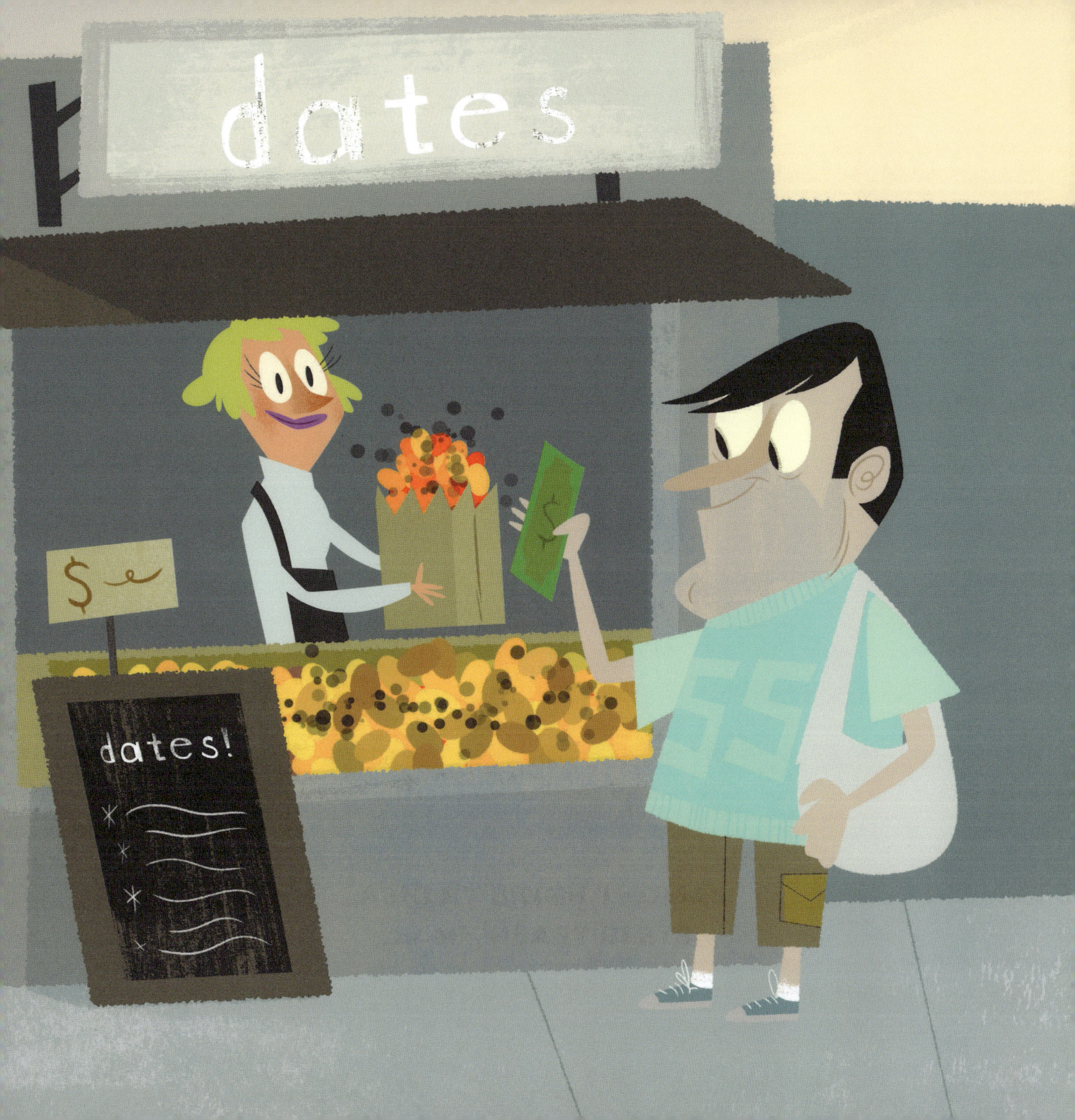
dates
dates!

Your mother shook her head at him.
You don't like dates.
No one in your family likes dates.
Why did your father buy those dates?

Finally, your mother fed the dates to your chickens.

Peck, peck, peck. They loved the dates!

A young hen laid an egg that had a little bit of carbon in its yellow yolk. You and your brother went out to gather eggs.

EGG FIGHT!

(I told you it was your brother's fault.)

Only two eggs were left for breakfast. You ate the egg with a little bit of carbon that used to be in a dinosaur.

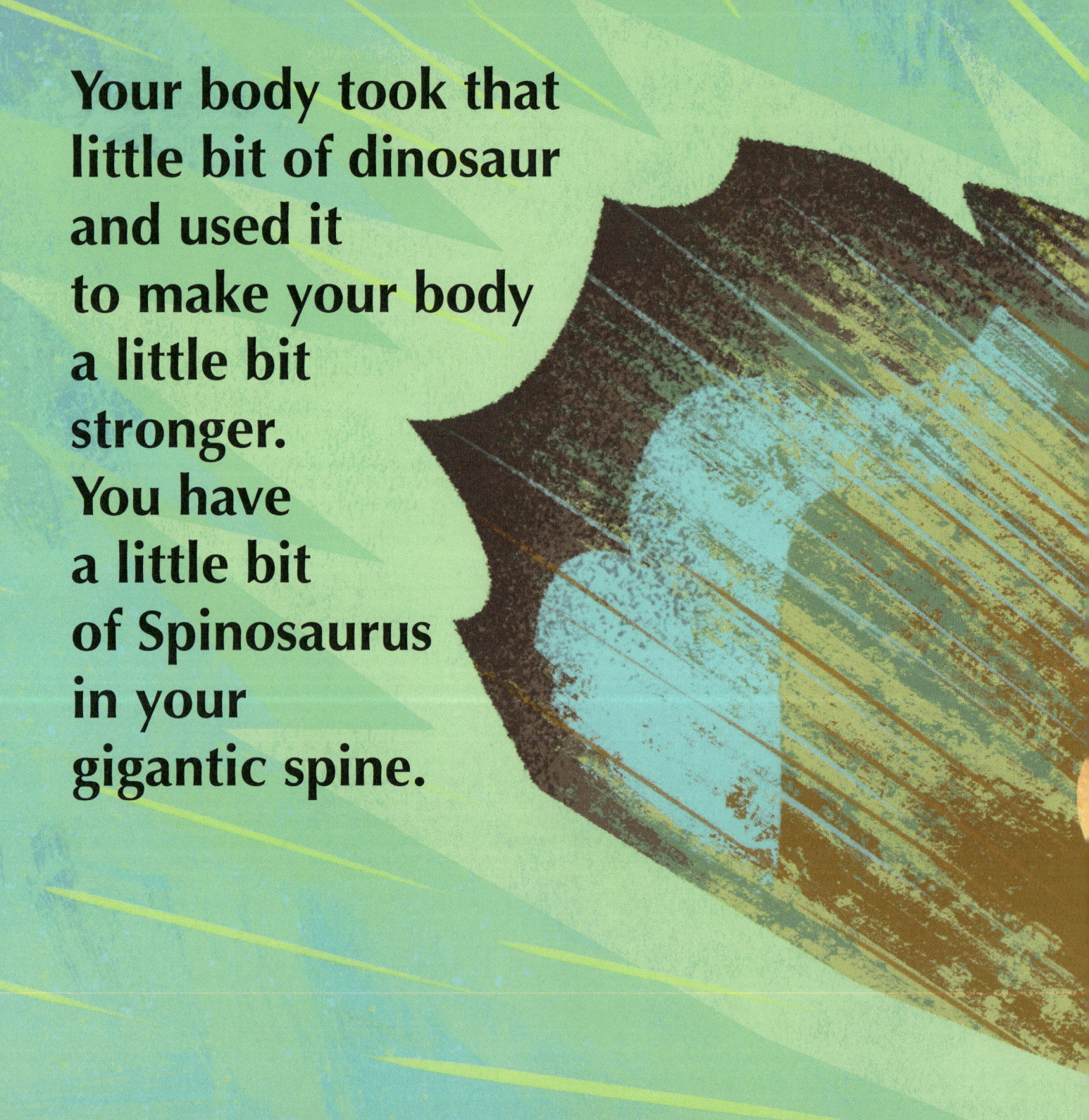

Your body took that little bit of dinosaur and used it to make your body a little bit stronger. You have a little bit of Spinosaurus in your gigantic spine.

If you look closely, you might see a little bit of Utahraptor in your fierce eyes, OR…

a little bit of diplodocus in your tailbone, OR...

...a little bit of triceratops on your forehead, OR...

a little bit of iguanodon in your fast legs. And when...

...your days on earth are done, and your body returns to the land, that little bit of dinosaur will be used again.

Maybe someday the carbon will **travel** across continents and wind up in the scales of a cutthroat trout.

The Carbon Cycle

All matter and all living things are made up of atoms, tiny particles that you can only see with powerful microscopes. The main atoms that make up living thing are carbon, hydrogen, oxygen, nitrogen, phosphorous, and sulfur. When any living thing dies, the atoms that make it up are used over and over in new living things. The atoms go from a living thing to the soil, air, or water, and then back into a living thing.

Carbon in the air starts the carbon cycle. The carbon combines with oxygen to make carbon dioxide. Plants use the carbon dioxide to store food in their leaves, roots, or fruit. When animals eat the plants, the carbon atom becomes part of the animal, maybe bones or muscle.

Maybe they are eaten by another animal.

When the animal dies and the body decomposes and rots away, the carbon atom returns to the soil, air, or water. Then the cycle starts all over again. Another plant can use the carbon atom, then another animal can eat the  plant, and the carbon becomes part of that animal. The cycle can repeat endlessly.

Printed in Great Britain
by Amazon

56757338R00021